MOLECULES

Rebecca Woodbury, Ph.D., M.Ed.

Gravitas Publications Inc.

MOLECULES

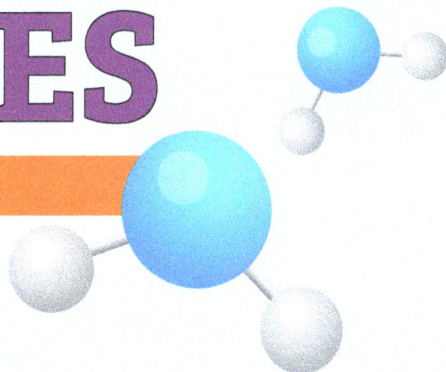

Illustrations: Janet Moneymaker
Design/Editing: Marjie Bassler, Rebecca Woodbury, Ph.D., M.Ed.

Molecules
ISBN 978-1-950415-10-6

Published by Gravitas Publications Inc.
Imprint: Real Science-4-Kids
www.gravitaspublications.com
www.realscience4kids.com

RS4K

Image credits: Cover & Title Page, By artegorov3@gmail, AdobeStock , Above, iStock: Macrovector

We know that everything is made of **atoms**.

Those are atoms!

Review

Atoms are tiny building blocks that can link together.

Atoms make everything we see, touch, taste, and smell.

We can draw atoms with arms and hands to help us learn how they work.

Hydrogen

Oxygen

Carbon

Atoms can link together
to make **molecules**.

Look! A salt
molecule!

Sodium

Chlorine

Sodium chloride
(table salt molecule)

Molecules can be **small** like salt, water, and methane.

Salt molecule

Methane molecule

Water molecule

Molecules can be **big** like sugar.

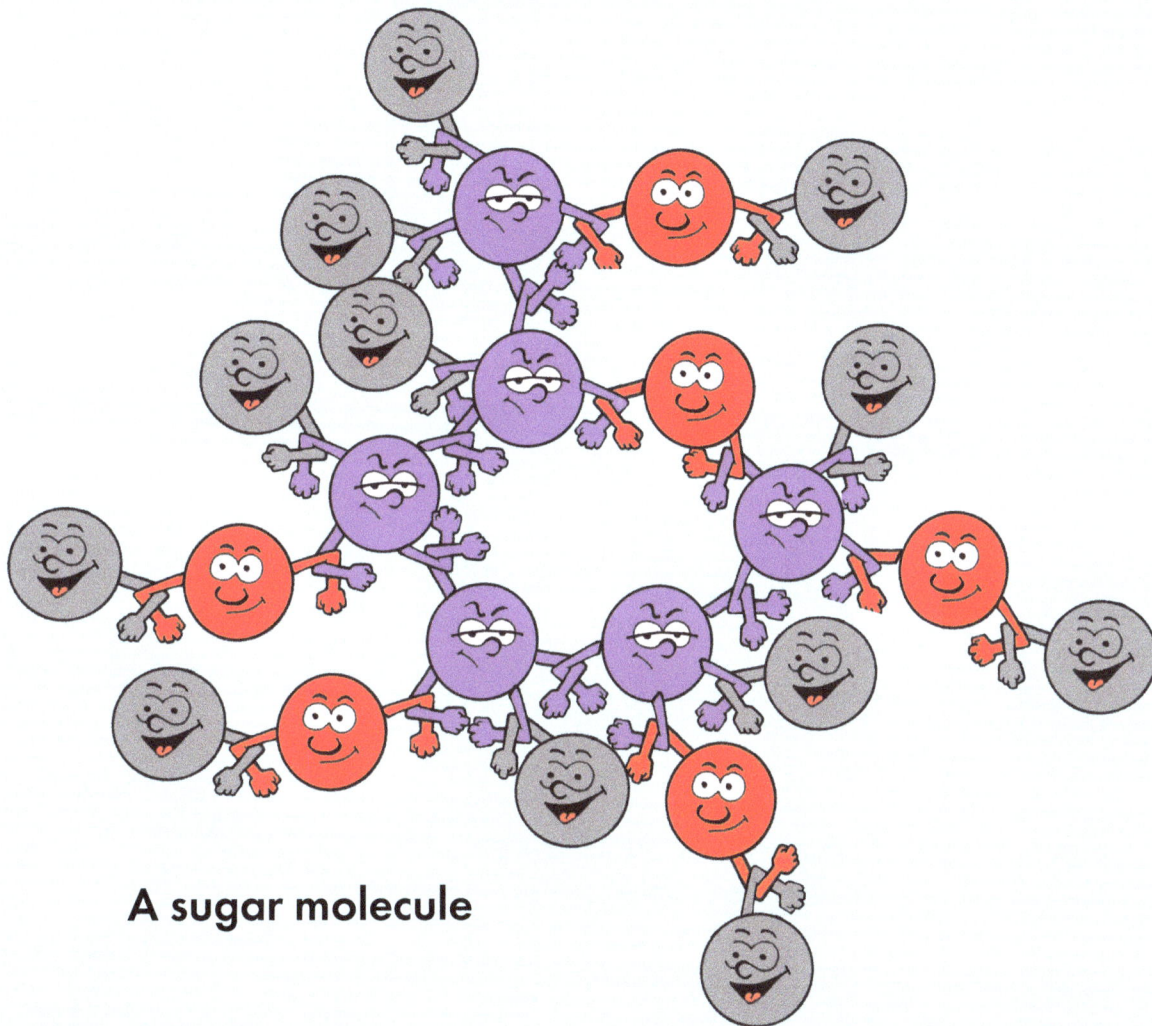

A sugar molecule

Molecules can be really big and made of many atoms!

A **protein** is a molecule made of thousands of atoms.

Look at all those atoms!

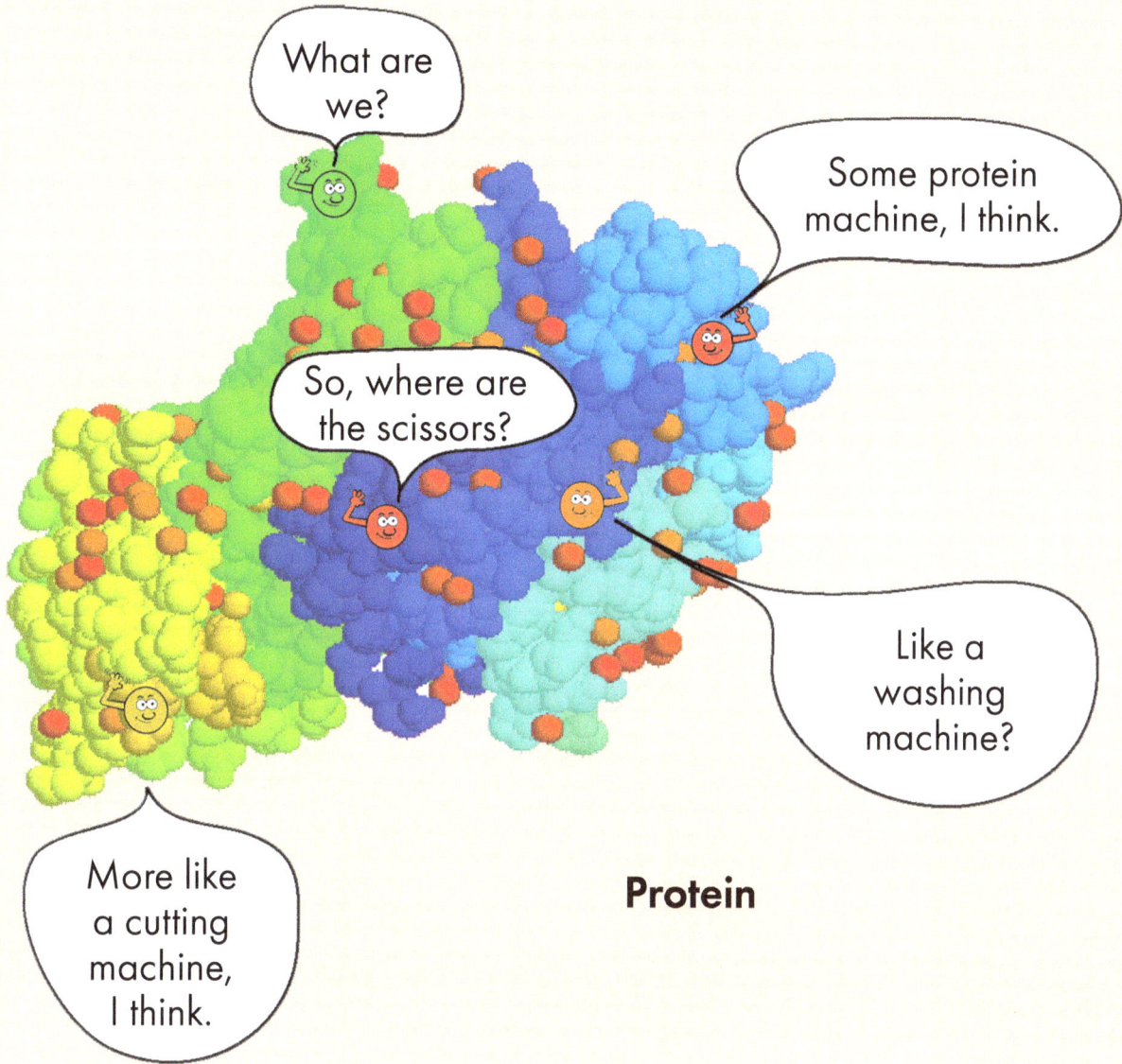

Protein

Atoms link together with **electrons**.

Electrons are shown as arms in the drawing of atoms.

We can call the linking arms **linking electrons**.

Could we link arms to make a molecule?

Probably not.

Our arms are linking electrons.

Sodium   Chlorine

Linking electrons
make a bond.

Sodium chloride
(table salt molecule)

Atoms have to follow **rules** when they make molecules.

Rules

1. **Atoms** link using their **linking electrons**.

2. **Atoms** form a **bond** when they link.

3. The number of **linking electrons** **equals** the number of **bonds**.

I have two arms.
I make two bonds.

Atoms

Hydrogen

Oxygen

Hydrogen

We follow
the rules!

Water molecule

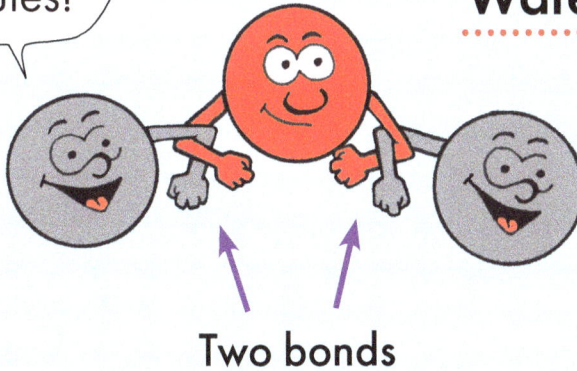

Two bonds

A hydrogen atom has **one arm**.

A hydrogen atom has only **one linking electron**.

Hydrogen

A hydrogen atom must follow the rule to make a molecule.

Rule

A **hydrogen atom** can form **only one bond**.

We can only make one bond.

One arm and no legs!

Hydrogen molecule

An oxygen atom has **two arms**.

An oxygen atom has **two linking electrons**.

Oxygen

An oxygen atom must follow the rule to make a molecule.

Rule

An **oxygen atom** can form **two bonds**.

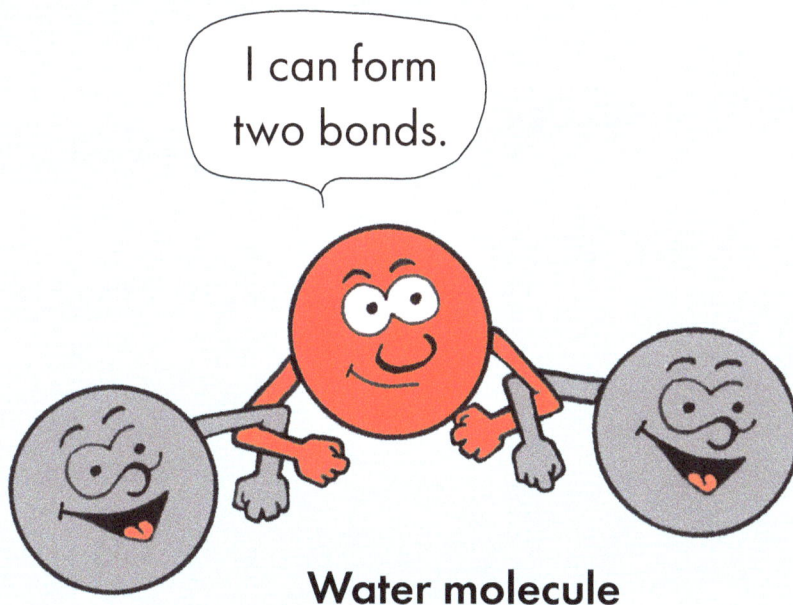

I can form two bonds.

Water molecule

A nitrogen atom has **three arms**.

A nitrogen atom has **three linking electrons**.

A nitrogen atom must follow the rule to make a molecule.

Rule

A **nitrogen atom** can form **three bonds**.

How many arms does carbon have?

How many linking electrons does carbon have?

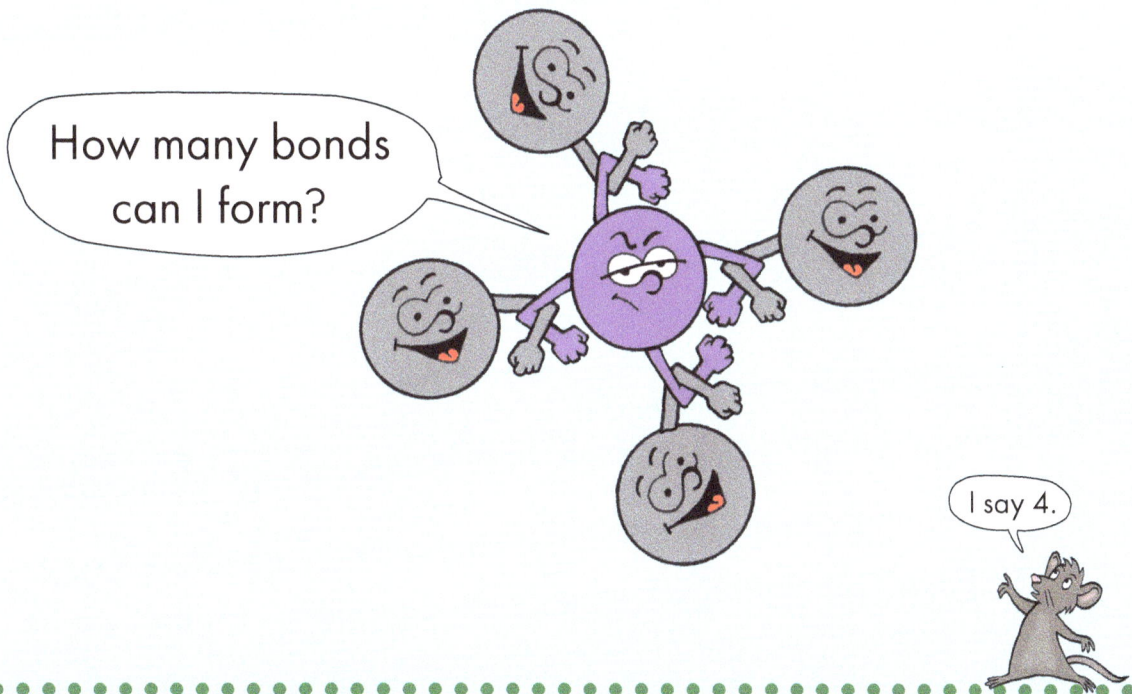

How to say science words

atom   (AA-tum)

carbon   (CAR-bun)

electron   (i-LEK-trahn)

hydrogen   (HIY-druh-jun)

methane   (MEH-thain)

molecule   (MAH-lih-kyool)

nitrogen   (NIY-truh-jun)

oxygen   (OCK-si-jun)

protein   (PROH-teen)